Alan Thomas and Pe

Avoca
Our Mining Heritage
A brief history of metal mining in the
Vale of Avoca, County Wicklow

Published by the Geological Survey of Ireland

The Vale of Avoca as seen from Tigroney (East Avoca)

Sponsorship provided by Ballykissangel

Design and Layout by Cartography Unit

Line drawing by Robert J. Tyrrell

Printed in Ireland by Betaprint

ISBN 1 899702 16 4

© Geological Survey of Ireland 1998

All rights reserved. No part of this book may be reproduced by any means without prior permission of the Director, Geological Survey of Ireland, Beggars Bush, Haddington Road, Dublin 4. Tel. (01) 6707444 Fax (01) 6681782

METAL MINING IN THE VALE OF AVOCA

The Vale of Avoca is a beautiful part of County Wicklow, deservedly known as the "Garden of Ireland". It features on the earliest known map of Ireland by the geographer Ptolemy who is said to have visited the area in 150AD. The mineral wealth of the valley has been known for centuries and prominent people have been involved in its exploration. Among the most celebrated (though not the most successful) was the 19th century nationalist leader Charles Stuart Parnell who was said to be obsessed by exploration progress. The Vale's beauty has been praised by the poet Thomas Moore (1779-1852) with lines such as :-

There is not in the wide world
a valley so sweet
As the Vale in whose bosom the
bright waters meet
Oh! The last rays of feeling
and life must depart
'Ere the bloom of that valley
shall fade from my heart

An eminent geologist Sir William W Smyth visiting the area in 1853 wrote of his impressions -

There is perhaps no tract in these islands which exhibits, even to the eye of the uninitiated, an appearance so strongly stamped with the characteristics of the presence of metallic minerals. For a considerable distance on both sides of the deeply cut valley of the Ovoca the face of nature appears changed and instead of the grassy or wooded slopes, or the grey rocks which beautify the rest of its course, we see a broken surface of chasms, ridges and hillocks, glowing with tints of bright red and brown, or assuming shades of yellow or livid green, which the boldest artist would scarcely dare to transfer to his canvas.

Here and there from among the ruins peers the white stack and house of a steam engine; or water wheels stand boldly

View towards the Williams Engine House at Tigroney (East Avoca), with old mine dumps on the lower ground.

projected against the hillside, some still and neglected, others whirling around in full activity; long iron pump rods ascend the acclivities to do their work at distinct shafts, and as long as daylight lasts, the rattle of the chains for raising the ore, and the clink of the separating hammers attest the vigour of the operations.

In truth quite independently of the geological or mining interest of the place, a walk through this series of mines, especially on a sunny evening, will yield a harvest of novel and striking scenes, the effect partly of the features of the mineral ground and partly the fine distant prospects which the higher workings command.

Today the beauty of the valley has reached a much wider audience through the success of the BBC-TV series "Ballykissangel" which is filmed on location in Avoca.

THE HISTORY OF MINING

It is likely that iron mining has taken place for many centuries in the wider Avoca district. However in the Vale itself the first recorded mine production took place around 1720. Copper was mined in the mid 1700s in both East (Cronebane) and West (Ballymurtagh) Avoca. At the same time some silver and a little gold was won from the oxidised cap (gossan) overlying the copper ore at Cronebane. A silver flagon made from this silver is still in existence.

In 1778 the Cronebane Company started mining for copper at Cronebane and Tigroney (East Avoca). The men involved had been the founders of the famous Parys Copper Mine in Anglesea (Wales) but had lost the lease to it. In 1790 Abraham Mills of the Cronebane Company joined with other owners to form a company which was incorporated by Act of Parliament in 1791 as the Associated Irish Mine Company. This company extracted copper ores from the mines and at its

Nineteenth Century Engine houses and additional 1940s old workings at Ballymurtagh, West Avoca.

peak employed up to 2000 people in the valley. The manager, Thomas Weaver, vigorously developed the mines and, in addition, copper was recovered from mine waters through the use of scrap iron. Production at the mines was halted temporarily by the 1798 Rebellion during which peasant insurgents were defeated by British forces including members of the mine management (who were subsequently rewarded for their "spirited and judicious conduct").

In 1811 further links with copper mining in Cornwall were established when the Williams Brothers of Perran-ar-Worthal, Cornwall acquired the lease of Cronebane and Tigroney mines, which their company held for the next seventy years up to 1884. During the period 1840-1880 pyrite for the manufacture of sulphur became a premium commodity of the Avoca mines and once more some 2000 people were employed in the mines of the valley. Copper ore was also mined but was a by-product of the pyrite at this time,

Artist's impression of the Avoca valley mines in the mid 1860s : Ballymurtagh, West Avoca, viewed from East Avoca.
Robert J. Tyrrell Artist

a situation which was not reversed for another hundred years when copper again became the primary product in the modern mines. The Williams Brothers, together with Henry Hodgson at Ballymurtagh and Ballygahan in West Avoca, introduced many improvements in the mining methods. Steam was used to power the water pumps, hoists and crushing equipment. Hodgson's tramway was built to transport ores to Arklow port for shipment to Swansea smelters. In spite of this profit declined steadily up to the 1880s when larger scale mining ceased.

Intermittent and small scale production, mainly for pyrite, took place in the decades that followed. The potential for pyrite and copper ore was explored by the State owned mining company Mianraí Teoranta which led to the establishment of large, low grade ore reserves in 1956. This in turn brought in the Canadian mining company Mogul International to mine some three million tonnes of copper and

This archway carried Hodgson's Tramway at Ballymurtagh, West Avoca.

pyrite ore, trading as St Patrick's Copper Mines Ltd. This company was forced by falling ore grades and metal prices to close in 1962. The actual closure was shrouded in some mystery as it followed the loss of an ore carrier off the Irish coast.

A period of care and maintenance followed until 1969 when a consortium led by Discovery Mines of Canada reopened the mine as Avoca Mines Ltd. By use of trackless mining methods this company mined some 8 million tonnes of ore until it too was forced by falling grades and metal price to close in 1982 and the underground workings are now flooded.

AVOCA GEOLOGY

Why was Avoca so richly endowed with mineral deposits as to give rise to its extensive mining heritage? The answer lies in the fact that its bedrock consists of volcanic ashes which developed in an environment which was favourable for the evolution of mineral deposits. Avoca lay close to the boundary between two colliding plates at this time - about 450 million years ago! The volcanic ashes accumulated on a seafloor where jets of metal-rich superheated waters were injected into the cold seawater, causing the metals to precipitate out as a thin blanket of sulphides. This chemical activity altered the composition of the volcanic ashes containing metal deposits and therefore this dark green (chlorite-rich) ash is a reliable guide to where such deposits may be found.

Copper occurs mainly as chalcopyrite, or fool's gold, composed of copper, iron and sulphur. The ore typically contains 1% copper or less. Pyrite, an iron-sulphur mineral valued for its sulphur content, was originally used for making gun powder but more recently has been used in fertiliser production. Lead (in the form of galena) and zinc (as the mineral sphalerite) occur with the copper in minor quantities but have not been recovered commercially. Ochre, naturally occurring yellow-brown pigment consisting of iron oxides and kaolin, forms the near-surface expression of the copper and pyrite deposits and was mined on a small scale until the 1940s.

View of the Footwall of the Cronebane Open Pit showing the typical green colour of the volcanic host rocks for the ores. The upper part of the pit face is brown due to the oxidisation of the minerals present originally. In the foreground the floor of the open pit has been backfilled with mine waste.

MINING METHODS

Although the Vale of Avoca has produced a wide range of minerals and metals over the centuries, copper has been the most important, having been mined from open pits and from underground since about 1720. The mining of the ore from underground requires efficient machinery to keep the mine free of water, to crush and hoist the broken ore to the surface and subsequently to treat the ore to produce grades which can be sold to overseas smelters to produce the copper metal.

The artist's impression of the valley's mines gives some idea of the layout of the machinery. The ore bodies in the Avoca valley are sheet-like in shape, 2-20 metres thick and are steeply disposed. In the early mining operations the underground ore was reached by tunnels (adits) driven into the sides of the valley. To allow for the natural drainage of mine water they were driven at a slight gradient up hill from the entry. Once the ore was reached the object was to extract it with the minimum

Artist's impression of the Avoca valley mines in the mid 1860s : Tigroney, East Avoca, looking across from West Avoca.
Robert J. Tyrrell Artist.

Underground mine operations showing typical working methods and style of dress in the late nineteenth century. Taken from an original photograph by J. C. Burrows, Camborne, Cornwall.

of waste rock and leaving the rock walls which contained the orebody intact.

Water entering any mine is always a problem. With modern mines electric pumps can now handle large quantities of water from great depths below the surface. In the 1700s and 1800s however, pumping water from below adit level was an expensive and difficult task and limited the depth of the mine and thus its productive life. The method of driving the underground tunnels and breaking the ore consisted of men drilling holes in the ore using a drill and sledgehammer - hard exhausting work by the light of a tallow candle. Once the series of holes was completed they would be charged with gunpowder and exploded to break the rock into sizes which could be hand shovelled into wheelbarrows or small track-mounted wagons which were then pushed out along the adit to the surface.

In the case of mine working below adit level, rock would be wheeled to a shaft and dumped into a steel bucket attached to the surface hoist by rope or chain. Before steam was used the hoist would have been worked by a horse walking in a circle attached to the drum which held the rope or chain.

Once on surface the ore had to be treated to remove waste rock and other impurities so that a product containing an economic quantity of copper and/or pyrite could be sold to the smelter at a profit. Ore was washed, picked to remove waste material, broken into uniform sizes by hand - at each stage the best copper was stockpiled for sale to the smelter. Poorer quality ore would be further reduced in size so that the best grade could be recovered. The ore for sale to the smelters - usually in Wales - was taken by horse-drawn carts to the port in Arklow. From 500-1000 cart-loads per day were sometimes needed to haul the ore to the ports. At the ports the ore was loaded onto ships for transfer to larger ships in Dublin or Dun Laoghaire for shipment to Swansea. During harvest time horses for drawing the ore to the ports would be scarce as they were needed on the farms. To solve this problem and reduce transportation costs Henry Hodgson built a mineral tramway from west Avoca to Arklow port. This enabled him to transport his ore by rail, initially with horse-drawn wagons and later using steam locomotives to haul the wagons.

Hodgson's steam engines and wagons were eventually sold to the Dublin-Wexford railway company in 1861.

By the mid 1800s the use of steam engines to provide power for hoisting, pumping and crushing in the mines of Cornwall was widespread and many design improvements to the engines had been made by the Cornish mine engineers. The Williams Brothers and Henry Hodgson soon took advantage of these Cornish engines and installed them in their mines at Avoca. Today only the empty buildings which once housed these great steam engines stand with their tall stacks, like sentinels guarding the valley's industrial and mining heritage.

Water wheels were also used at Ballygahan at river level and at Tigroney for driving pumps and stamps. Some wheels were 50ft in diameter. Flat rods of iron attached to the water wheels enabled power to be transmitted up hill to pumps situated in the various shafts, thus saving the expense of operating the more costly steam engines in the shallower workings.

The Ballygahan headframe in 1959, West Avoca.

Interior of the mill at Ballygahan, West Avoca. About 1960.

MODERN MINING (1958-1982)

Mining in this period in the Avoca Valley relied for the greater part on underground trackless mining - extracting large tonnages of lower grade ore in the West Avoca area at Ballygahan and Ballymurtagh. Open pit mining was mostly from the east side of the river at Tigroney and Cronebane. From 1958-1962 St Patrick's Copper Mines Ltd. was responsible for the introduction of the relatively new trackless method of underground mining which enabled larger tonnages to be mined and consequently lower grades could be economically extracted. This entailed driving an inclined tunnel from surface down at a gradient of 1 in 8 to gain access to the ore body. Simultaneously a steeper tunnel driven from surface to an underground crusher station was put in to house a series of four

Visitors to Avoca in the 1950s (left to right):
Dr M. A. Hogan, Chairman, Mianraí Teoranta,
Mr. A. Woods, Director, Mianraí Teoranta,
Mr. Sean Lemass, Tanaiste and Minister for Industry and Commerce (subsequently Taoiseach) and
Mr. T. R. H. Nelson, Mine Manager, Mianraí Teoranta.

conveyor belts which took the crushed ore from the underground storage bins under the crusher station and conveyed it to the surface.

Then it went into the mill ore storage bins. Ore from these surface bins fed the newly constructed ore processing plant or mill as it is known. This mill further reduced in size, screened, washed and crushed to a fine powder. Each stage produced a separation of higher grade from waste material. The difference being that now none of these operations was done by hand, but by large electrically driven machines. The use of chemical reagents was also a 20th century innovation as flotation of the ore

Drilling operations underground at west Avoca in the 1970's. Note the scale of the tunnels in relation to the miners on the drilling machine.

was capable of handling 4000 tonnes of ore per day. Once in the mill the ore was treated as it was in the 19th century but using modern equipment; it was to cause the needed separation was aided by the addition of a variety of reagents, depending on the impurities which had to be separated from the copper concentrate. Copper concentrate powder with some 20% contained copper was the principal product at the mine with pyrite being a secondary product sold to the local fertiliser factory for the manufacture of sulphuric acid. The copper concentrate was trucked to Arklow port for shipment to European smelters for the production of copper metal.

Avoca Mines Limited, after some re-furbishment of the existing plant and using the more modern low profile mining equipment, continued to develop and extract the ore from the West Avoca underground workings. This led to the mine being deepened and a new underground crusher and conveyor being installed to improve productivity. Ore was mined from two main ore bodies, South Lode and Pond Lode. Improved stoping methods developed at the mine at this time enabled greater output to be achieved from the two ore bodies than had previously been possible. One of the techniques developed here has since become known internationally as the Avoca Method. The company also mined ore from open pits at East and West Avoca.

The spoil heaps from 18th and 19th century mining were re-treated using the modern processing techniques which made a better recovery of copper possible. What had been thrown aside as waste in the past was now capable, using these modern techniques, of being transformed into a saleable product. Sadly history was to repeat itself and in spite of financial support from the Government, the company once again was forced into receivership on 6 August 1982. Once the underground pumps were switched off the mine started to fill with water and is now completely flooded. Thus ended another chapter in the centuries old metal mining history of the Vale of Avoca.

Attention is drawn to the fact that there are open workings and derelict buildings on these lands and extreme care should be taken to avoid them.

Some of these properties are State-owned. Queries concerning access to them should be directed to the Exploration and Mining Division, Department of the Marine and Natural Resources, Beggars Bush, Haddington Road, Dublin 4.

Aerial view of the mill and mine workings at Ballygahan (left background) and Ballymurtagh (upper right), West Avoca. There is a tailings impoundment in the foreground. About 1956.

THE FUTURE OF THIS HISTORIC MINING AREA

With the ever increasing interest in industrial, cultural and mining heritage worldwide, the Vale of Avoca has a unique opportunity to become the premier Irish mining heritage centre. It is intended to nurture this mining heritage by initially preserving the 19th century Cornish engine houses on both sides of the valley. Also envisaged are a series of trails through the old mine areas once they have all been made safe for the public to explore. On the west side of the Avoca River there are plans for a Miners' Park and Museum, and it is intended to develop a separate Mining Heritage Centre on its east bank. Check whether either or both have opened, they should be worth a visit. We will also encourage renewed contact with our Celtic mining

Hodgson's tramway arch today.
Robert J. Tyrrell. Artist

neighbours in Cornwall and Wales.

We hope you enjoy your visit to the Vale of Avoca. Food and drink are available in many places, including The Meetings, The Old Coach House, Vale View Hotel, Fitzgeralds, Avoca Inn, Woodenbridge Hotel, Valley Hotel and Avoca Handweavers. More information on Avoca's mining history is contained in G.A.J. Cole's (1922) "Memoir on metalliferous localities", now out of print but available in many libraries and the Geological Survey of Ireland (GSI). Another GSI publication, "Enjoying the Vale of Avoca", describes its geology and is available locally or from GSI.

This guide aims only to provide a summary of Avoca's mining heritage: it does not give a right of entry into private property for which permission should always be sought from the owner. The State does not accept liability for any injury, loss or damage sustained by a person or caused to his/her property as a result of visiting any location described in the guide.

Back cover: Ore sample and inset showing the Baronets Engine House, East Avoca (1960s)